LIGHTNING BOLT BOOKS™

Let's Look at Brown Bears

Ruth Berman

Lerner Publications Company

Minneapolis

Lerner Publications Company
A division of Lerner Publishing Group, Inc.
241 First Avenue North
Minneapolis, MN 55401 U.S.A.

Website address: www.lernerbooks.com

Library of Congress Cataloging-in-Publication Data

Berman, Ruth.
    Let's look at brown bears / by Ruth Berman.
       p.   cm. — (Lightning bolt books™ – Animal close-ups)
    Includes index.
    ISBN 978–0–7613–3890–1 (lib. bdg. : alk. paper)
       1. Brown bear—Juvenile literature. I. Title.
    QL737.C27B4532 2010
    599.784—dc22   2008051855

Manufactured in the United States of America
1 2 3 4 5 6 — BP — 15 14 13 12 11 10

Contents

Brown Bears

These are brown bear tracks. How many toe marks can you count?

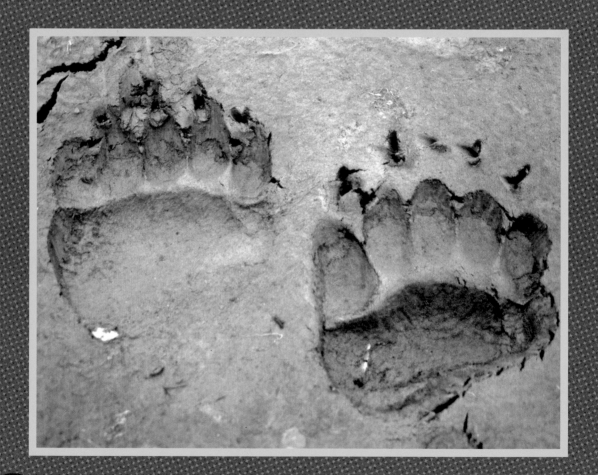

This is an Alaskan brown bear. Alaskan brown bears live near water.

Rivers make good wading spots for Alaskan brown bears.

They have small ears, small eyes, and a big, long nose.

Bears can stand up on their hind legs. This mother bear stands to smell the air for food and for enemies.

Bear Cubs

Mother bears usually have twins or triplets. Baby bears are called cubs.

Twins are common in bear families.

Cubs stay with their mother
for one to three years.

This cub is napping on Mom!

These cubs are playing.

Eating Habits

Brown bears are omnivores.
Do you know what an
omnivore is?

This brown bear is eating grass.

An omnivore is an animal that eats both plants and animals. Are you an omnivore?

Brown bears also eat fish.

Most furry animals walk on their toes. Bears walk with their feet flat on the ground.

Can you see this bear's toes?

This bear is walking on a trail. The trail ends at a river. Brown bears look for fish in rivers.

Look! These bears are fighting over a good fishing spot.

The bigger bear wins. It is about to catch a salmon!

Bears sometimes catch salmon as they jump out of the water.

What do you think this bear is doing?

It is fishing for salmon under the water.

This bear is pouncing on a salmon.

Bears move quickly to catch salmon.

This small bear is trying to sneak some food!

19

Alaskan brown bears
also eat clams.

Bears have to dig for clams.

This bear has found a clam to eat.

Getting Ready for Winter

Alaskan brown bears get ready for winter by eating a lot. Eating a lot makes them fat.

Why do bears need to get fat? Fat keeps them warm and healthy during the winter.

This bear stays warm on the coldest days.

Brown bears stay in dens for most of the winter. A brown bear is hibernating in this cozy den. When bears hibernate, they are in a deep sleep.

Springtime

In the spring, bears leave their dens to look for food.

A bear comes out of its den in the spring.

Then bears eat and eat. They will be fat again by next winter.

Two bears fish together.

Fishing is hard work!
It is time to rest.

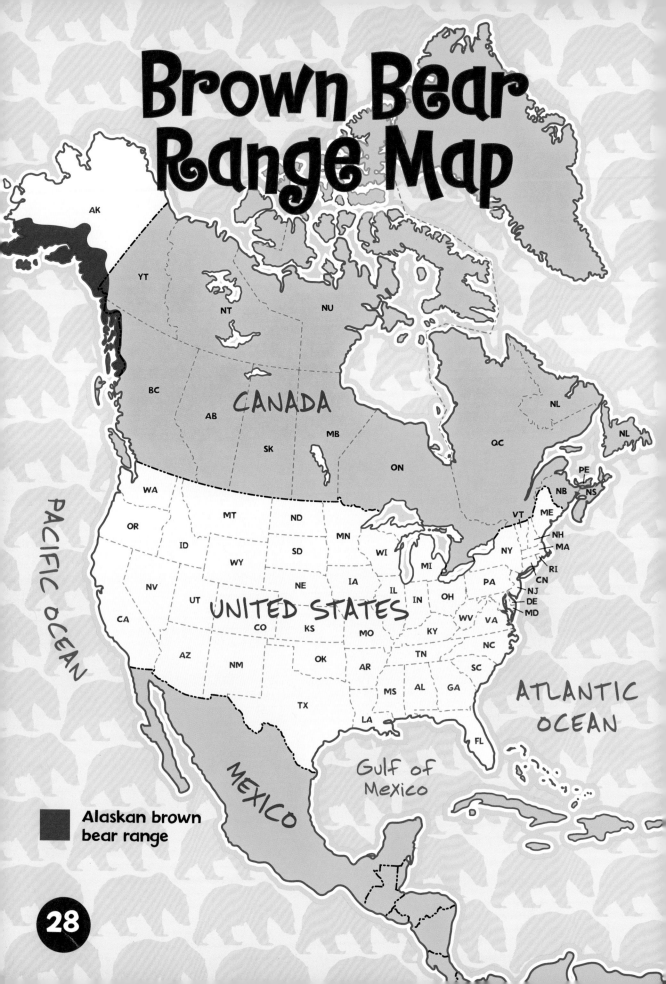

Brown Bear Range Map

Alaskan brown bear range

Parts of a Brown Bear's Body

hump
(strong shoulder muscle)

ear

eye

nose

mouth

front
legs

hind
legs

paw

claws

Glossary

clam: a boneless animal that lives inside a shell. Clams are found under wet sand or mud. The soft meat of a clam can be eaten.

cub: a baby bear

den: a cozy, safe place to live

hibernate: to spend most of the winter in a deep sleep

omnivore: an animal that eats both plants and animals

pounce: to jump onto something suddenly

salmon: a kind of large fish

triplets: three babies born at one time to the same mother

twins: two babies born at one time to the same mother

Further Reading

Bears at Enchanted Learning
http://www.enchantedlearning.com/themes/
bear.shtml

Berger, Melvin, and Gilda Berger. *Do Bears Sleep All Winter?: Questions & Answers about Bears.* New York: Scholastic, 2001.

Hodge, Deborah. *Looking at Bears.* Toronto: Kids Can Press, 2008.

Murray, Julie. *Grizzly Bears.* Edina, MN: Abdo, 2002.

National Geographic Creature Feature:
Brown Bears
http://kids.nationalgeographic.com/Animals/
CreatureFeature/Brown-bear

San Diego Zoo's Animal Bytes:
Brown Bear
http://www.sandiegozoo
.org/animalbytes/
t-brown_bear.html

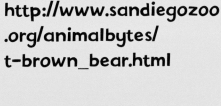

Index

Photo Acknowledgments

The images in this book are used with the permission of: © Photodisc/Getty Images, p. 1; © Tjommy-Fotolia.com, p. 2; © Gerard Fuehrer/Visuals Unlimited, Inc., p. 4; © Patrick J. Endres/Visuals Unlimited, Inc., p. 5; © James Urbach/SuperStock, p. 6; © John Warden/Riser/Getty Images, p. 7; © age fotostock/SuperStock, pp. 8, 16, 19, 21; © Michael S. Quinton/National Geographic/Getty Images, p. 9; © Norbert Rosing/National Geographic/Getty Images, p. 10; © Cary Anderson/Aurora/Getty Images, p. 11; © Klaus Nigge/National Geographic/Getty Images, p. 12; © Andy Rouse/The Image Bank/Getty Images, p. 13; © Johnny Johnson/Riser/Getty Images, p. 14; © Jeff Foott/Discovery Channel Images/Getty Images, p. 15; © Matthias Breiter/Minden Pictures, p. 17; © Johnny Johnson/The Image Bank/Getty Images, p. 18; © Vincenzo Lombardo/Taxi/Getty Images, p. 20; © Rick Parsons-Fotolia.com, p. 22; © Johnny Johnson/Photographer's Choice/Getty Images, p. 23; © STOUFFER PRODUCTIONS/Animals Animals, p. 24; © Shane Moore/Animals Animals, p. 25; © Oksanaphoto/Dreamstime.com, p. 26; © Steve Winter/National Geographic/Getty Images, p. 27; © Laura Westlund/Independent Picture Service, pp. 28, 29; © Sandrine Pinard-Fotolia.com, p. 30; © Roman Krochuk-Fotolia.com, p. 31.

Cover: © Andy Rouse/The Image Bank/Getty Images (main); © Daniel J. Cox/Photographer's Choice/Getty Images (mountain background).